Oxford International Primary Geography

5

Terry Jennings

OXFORD
UNIVERSITY PRESS

Great Clarendon Street, Oxford, OX2 6DP, United Kingdom

Oxford University Press is a department of the University of Oxford. It furthers the University's objective of excellence in research, scholarship, and education by publishing worldwide. Oxford is a registered trade mark of Oxford University Press in the UK and in certain other countries

British Library Cataloguing in Publication Data
Data available

978-0-19-831007-5

13 12 11

Paper used in the production of this book is a natural, recyclable product made from wood grown in sustainable forests. The manufacturing process conforms to the environmental regulations of the country of origin.

Printed in Great Britain by Bell & Bain Ltd, Glasgow.

Acknowledgments

The publishers would like to thank the following for permissions to use their photographs:

Cover photo: S-F/Shutterstock, P4a: Moment/Getty Images, P4b: ArabianEye/Getty Images, P6a: The Image Bank/Getty Images, P6b: Frank Krahmer/Corbis/Image Library, P7: Shutterstock, P8a: Jake Lyell / Care / Alamy, P8b: Shutterstock, P8c: ArabianEye/Getty Images, P10: Shutterstock, P12: Robert Fried / Alamy, P13a: Blaine Harrington III/Corbis/Image Library, P13b: Shutterstock, P16a: iStock, P16b: Gideon Mendel/Corbis/Image Library, P17: Afp /Getty Images, P18: Hugh Sitton/Corbis/Image Library, P19: Shutterstock, P20: Rajesh Kumar Singh/AP Images, P21a: Sean Sprague/ Robert Harding Picture Library, P21b: Charlotte Thege / Alamy, P21c: Sean Sprague / Alamy, P22a: Shutterstock, P22b: Shutterstock, P23a: Jacek Chabraszewski/ Shutterstock, P23b: Shutterstock, P23c: Shutterstock, P24: Brian Maudsley/Shutterstock, P25: Art Directors & TRIP / Alamy, P26: Photographer's Choice RF/Getty Images, P27a: Shutterstock, P27b: iStock, P28: Urbanmyth / Alamy, P29a: Shutterstock, P29b: Best View Stock/Getty Images, P30: Richard du Toit/Gallo Images/Getty Images,P31: Bob Daemmrich/Corbis/Image Library, P32a: Shutterstock, P32b: Photographer's Choice/Getty Images, P33: Shutterstock, P36: Ken Wolter/ Shutterstock, P37: Shutterstock, P38a: Tim Laman/National Geographic/Getty Images, P38b: Peter Clark/Shutterstock, P38c: Chris Minerva/ Photodisc/ Getty Images, P39: E+/ Getty Images, P40: John Borthwick/ Lonely Planet Images/ Getty Images, P41a: Dean Northcott/ Image Source/Corbis/Image Library, P41b: Andreas Pacek/Westend61/Corbis/Image Library, P42: Jonatan Martin/ Moment Open/ Getty Images, P43: Shutterstock, P44a: Georgette Douwma/The Image Bank/Getty Images, P44b: Danita Delimont/Gallo Images/Getty Images, P45: © travelib / Alamy, P46a: Tim Draper/Dorling Kindersley/ Getty Images, P47a: Getty Images News, P47b: Shutterstock, P48: iStock, P49: Afp /Getty Images, P50: Frans Lemmens/ Corbis/Image Library, P51: Lee Adamson / Alamy, P53a: Shutterstock, P53b: Shutterstock, P53c: Shutterstock, P54: Robert Hoetink/Shutterstock, P55: José Marafona /Dreamstime, P56: Nasa /Corbis/Image Library, P57: Anthony A FOSTER/AFP/Getty Images, P58a: Shutterstock, P58b:Toronto Star/Getty Images, P59: Getty Images Sport

Although we have made every effort to trace and contact all copyright holders before publication this has not been possible in all cases. If notified, the publisher will rectify any errors or omissions at the earliest opportunity.

Links to third party websites are provided by Oxford in good faith and for information only. Oxford disclaims any responsibility for the materials contained in any third party website referenced in this work.

Contents

We need water

Water is one of the most important substances on Earth. Without water to drink we would die. About 70 per cent of our body weight is water. People have lived for a month or more without food, but we could live for only three or four days without water.

How do we use water? We drink water. We wash our bodies, our clothes and many other things with water. We use water to flush the toilet. The foods we eat have water in them, and all our drinks are largely water. We use water to cook many of our foods, and we use water for leisure activities, such as boating, swimming and fishing.

Plants need water

Plants need water to live just like we do. Plants use their roots to take up water from the soil. The crops that farmers grow for our food need water to grow. In some places where there is little rain and the land is dry, farmers bring water from rivers, **lakes** and **wells** to their fields. The water is carried in pipes and ditches. This special kind of watering is called **irrigation**. Some crops need more water to grow than others. Rice and cotton, for example, need lots of water, while olives and oranges need only a little.

Farmers all around the world need water for their crops. This crop of alfalfa in Saudi Arabia is being irrigated.

We need water to keep ourselves clean.

Animals need water

Animals also need water to live. The animals that provide us with meat, milk and eggs need large amounts of clean water to drink. Water is also home to millions of plants and animals, including the fish and shellfish we eat.

Industry

Some **industries** use water to cool hot substances or the moving parts of machinery. Water also helps to produce most of our electricity. All **power stations** that burn fuels use water to make steam. This turns the machines that produce electricity. In hydro-electric power stations the water of fast-flowing rivers is used to turn the machines that produce electricity.

Water is used to make many kinds of food and many of the things we see or use every day.

How much water do animals and plants need?	
Human	2 litres per day
Dairy cow	135 litres per day
Large oak tree	20 000 litres per day

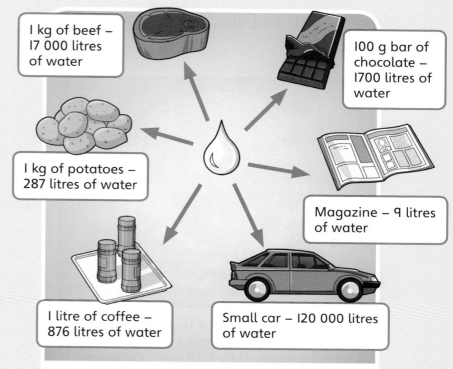

1 kg of beef – 17 000 litres of water

100 g bar of chocolate – 1700 litres of water

1 kg of potatoes – 287 litres of water

Magazine – 9 litres of water

1 litre of coffee – 876 litres of water

Small car – 120 000 litres of water

The amount of water needed to make or grow each of these things.

Did you know?

Earth is the only planet in the Solar System with liquid water.

Activities

1 Imagine you are going to a faraway planet. There is no water on the planet, so you will have to take supplies with you. Draw or write about what you will need water for.

2 a Do people in your area use water for leisure activities?

   b Write a list of the places where water is used for leisure activities.

   c Draw pictures to show the different places and activities.

Water everywhere

Look at a globe or the map of the world on pages 60–61. Can you see that nearly three quarters of the Earth is covered by water?

Nearly three quarters of the Earth is covered by water, but most of the water is salty seawater. Much of the freshwater is in the form of ice.

Salty water

About 97 per cent of all the water on Earth is salty. The water in the oceans and seas is always salty. Some of this salt has come from volcanoes under the sea, but most of it has come from rocks on land. When rain falls, it dissolves some of the salt in the rocks. Rivers and streams dissolve even more of the salt in rocks and carry it down to the sea.

Oceans

There are the five great oceans in the world. Can you name them? The largest is the Pacific Ocean. It covers about one third of the Earth and its average depth is over 4000 metres. All the oceans are connected to each other, so that their waters mix together.

Seas

Some seas, such as the Arabian Sea and the Sargasso Sea, are parts of oceans. Others, such as the Red Sea and the Black Sea, are surrounded by land, and so are separate from the oceans. The largest of the world's seas is the South China Sea.

Freshwater

We can only use freshwater for drinking, washing, cooking or watering plants. Most of the freshwater we use comes from lakes, rivers and wells.

Some freshwater is found in the form of ice. Most of this ice is around the North and South Poles, in huge sheets called polar ice caps. Ice is also found on the tops of high mountains. Sometimes ice flows down the mountain like slow-moving rivers. These are called **glaciers**.

There are huge sheets of ice, called polar ice caps, around the North and South Poles. This picture shows part of the Antarctic polar ice cap.

There is a lot of freshwater under the ground, which is called groundwater. Every time it rains, some of the water soaks into the ground through tiny cracks in the rocks. It is this groundwater that we reach when we get water from a well. Sometimes groundwater flows as an underground river, which may later come to the surface.

There is also a small amount of water in the air. Much of it is the invisible gas called **water vapour**. When water vapour in the air cools and turns into tiny droplets of water, it forms the clouds that we see.

Do you think this lake in Austria is freshwater or saltwater?

Did you know?

Although there is lots of water on Earth, only about 3 per cent of it is freshwater that we can use.

Activities

1 Work with a partner. Use the map of the world on pages 60–61 or an atlas to help you to answer these questions:

   a What is the name of the ocean off the east coast of Africa?

   b What is the name of the sea to the north of Libya and Egypt?

   c What is the name of the sea between Japan and North Korea?

   d What is the name of the ocean to the south of Australia and New Zealand?

   e Write the names of three seas that have colours in their name.

   f Which oceans and seas would a ship cross if it sailed from Lima in Peru to Muscat in Oman by the shortest route?

2 a Draw a plan of your school and its grounds.

   b Walk around the school and label your map with all the places where water moves, such as taps, drains, gutters and downpipes.

In one year the average family of four living in a town or city uses about 180 000 litres of water – enough to fill four large tanker trucks to overflowing. If we are to stay healthy, this water has to be clean and safe to use. In areas where it is difficult to get water, many people have to make do with only a few litres of water a day.

We store water by building dams across rivers. The water behind the dam collects, forming an artificial **lake**, or reservoir.

In many parts of Africa, children fetch their family's water from a **waterhole**, which may be up to 12 kilometres from their home.

How is water made safe to drink?

Water taken from a river or reservoir is dirty. The diagram opposite shows how it is cleaned and made safe.

Sewage

Used water is called sewage. Usually sewage is passed through big pipes to a **sewage** works for cleaning. Sewage contains lots of harmful germs and poisonous substances. It has to be cleaned before it can be safely returned to a river or pumped into the sea.

Where our water comes from

The water we use came originally from rain (or snow). In the country, some people still get their water from a stream or well. In towns and cities, streams and wells cannot provide enough water for the people who live there. In a few countries, freshwater is obtained from seawater. The salt is removed from the seawater by a process called **desalination**. In other areas, water is taken from lakes, rivers or **reservoirs**.

The world's largest desalination plant is the Jebel Ali Desalination Plant in the United Arab Emirates.

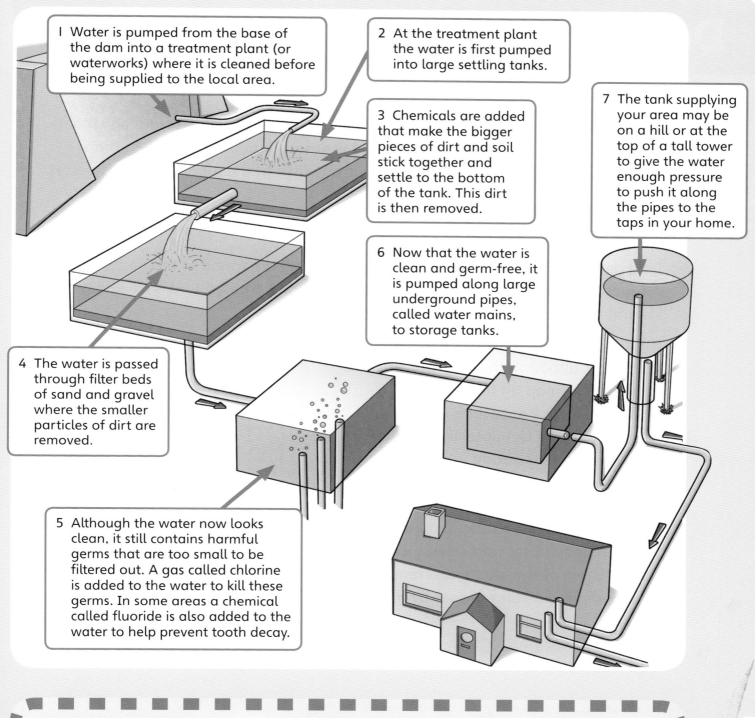

1. Water is pumped from the base of the dam into a treatment plant (or waterworks) where it is cleaned before being supplied to the local area.

2. At the treatment plant the water is first pumped into large settling tanks.

3. Chemicals are added that make the bigger pieces of dirt and soil stick together and settle to the bottom of the tank. This dirt is then removed.

7. The tank supplying your area may be on a hill or at the top of a tall tower to give the water enough pressure to push it along the pipes to the taps in your home.

6. Now that the water is clean and germ-free, it is pumped along large underground pipes, called water mains, to storage tanks.

4. The water is passed through filter beds of sand and gravel where the smaller particles of dirt are removed.

5. Although the water now looks clean, it still contains harmful germs that are too small to be filtered out. A gas called chlorine is added to the water to kill these germs. In some areas a chemical called fluoride is also added to the water to help prevent tooth decay.

Activities

1. If a bucket holds 10 litres of water, how many buckets of water does the average family of four use in one week? And in one year?

2. Imagine you had to collect all the water for your family from a well at the other end of your street. How would this change life in your home?

3. How is sewage in the sea harmful? Use reference books or the Internet to help you find out.

Deserts

Deserts are the driest places on Earth. Very little rain falls in deserts and few plants can grow. Scientists usually say that an area is desert if an average of less than 250 millimetres of rain falls each year. About one fifth of the land on Earth is desert.

Kinds of desert

There are two kinds of desert.

- Cold deserts: These deserts are dry and cold. They are found in the Arctic and Antarctic regions around the North and South Poles. The ground is covered with ice and snow for all or most of the year.

- Hot deserts: These deserts are dry and hot in the day, although they can be very cold at night. The map on this page shows where the main hot deserts of the world are found. Some hot deserts are sandy. The wind blows the sand into hills called **dunes**. However, in most hot deserts, the wind has blown the sand away, leaving pebbles or even bare rock. Sometimes there are mountains and steep rocky slopes. Some of the mountains have been carved into interesting shapes by the wind.

The world's largest hot desert is the Sahara Desert in North Africa. It has many large **sand dunes.**

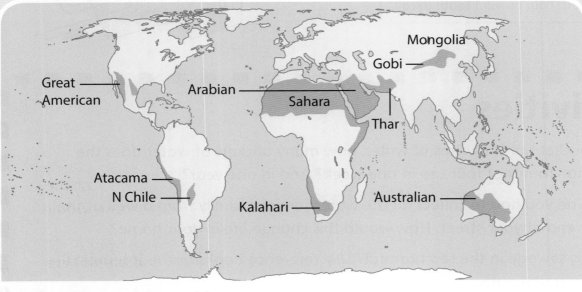

The hot deserts of the world.

Desert plants

Not many plants can survive in the desert. Those that do live there have special **adaptations** that allow them to survive the dry conditions. Some desert plants have very long roots to reach water underground. The roots of the mesquite bush of North America may go down 50 metres to reach water. A cactus stores water in its thick stem.

Water stored here

Cactus plants have many adaptations for surviving in a hot desert

Desert animals

Few animals live in the desert. Many of those that do get the water they need from their food. Most of them sleep during the day and only come out at night when it is cooler.

The best-known desert animal is the camel. A camel can go a long time without water. It can lose about one third of the weight of its body and still live. Then, when the camel does find water, it can drink 115 litres or more in a few minutes.

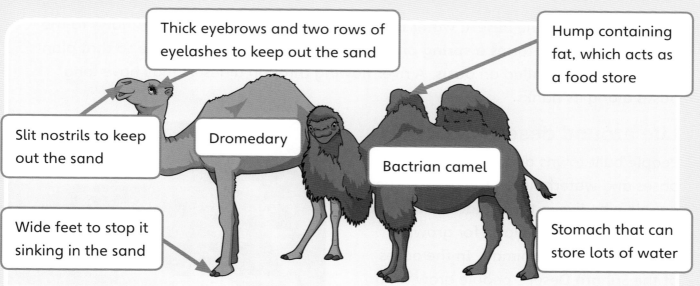

Thick eyebrows and two rows of eyelashes to keep out the sand

Hump containing fat, which acts as a food store

Slit nostrils to keep out the sand

Dromedary

Bactrian camel

Wide feet to stop it sinking in the sand

Stomach that can store lots of water

Different camels have different adaptations for surviving in different deserts. These camels live in different deserts. Which are they?

Activities

1 You are going to travel across the Sahara Desert or Arabian Desert. List the things you will take with you, in order of importance. What problems would you have in the desert? How would you cope with them?

2 Write and illustrate a book or make a class display on plant and animal life in deserts. Use reference books or the Internet to help you.

Life in the desert

In most parts of the world there is a lot of water under the ground. This is known as **groundwater**. Each time it rains, some of the water seeps down into the ground where it fills all the cracks and openings in the rocks below. Often it is possible to dig a well or borehole to reach this water.

Water in the desert

In deserts, most of the water from rain is lost. It either runs off the hard, dry surface or **evaporates** into the air. A little of it soaks into the ground where it may eventually join underground rivers and streams. Some of these underground rivers and streams bring their water from mountains hundreds of kilometres away, where there is heavy rainfall or snow.

In a few places in the desert, water from underground rivers or streams flows to the surface. The water forms a **spring** or **waterhole**, making the soil **fertile** so that plants can grow. This is called an **oasis**. A river flowing through a desert may have long oases along its banks.

Life around desert oases

People built towns and villages around oases and waterholes in the desert. The people who live in these places make use of every bit of fertile land for growing crops and rearing animals. In the oases of the Sahara Desert, people grow food plants such as vegetables, dates, figs, olives and apricots. They keep camels, sheep and goats for their meat and milk.

In this oasis in the Sahara Desert, people are able to grow crops by **irrigating** them with groundwater pumped from a well.

Nomadic people

Not all desert people live in oasis towns and villages. **Nomads** or wanderers also live in deserts. Some of the Bedouin of the Sahara and Arabian Deserts are nomads. They travel from one oasis to another, with their herds of sheep, goats and camels. The animals feed on the scattered desert plants. It is so hot during the day that the Bedouin often rest during the daytime and travel at night when it is cool. The nomadic Bedouin have no fixed homes. Instead, they live in tents made from goat skins or camel hair. To protect themselves from the heat, cold, and wind-blown sand, they wear long flowing robes.

Some of the Mongol people of the cold Gobi desert in Asia are also nomads. The nomadic Mongols are herders, keeping horses, sheep, cattle, goats and camels. They ride short-legged, long-haired horses when they round up or drive their herds of animals.

These nomadic Bedouins are in the Wadi Rum Desert.

This Mongol encampment is in the Gobi Desert in Asia. The round tents are called *gers*. They are made from thick felt stretched over a wooden frame.

Activities

1 Nomads do not build permanent homes. Write down why you think this is.

2 Use reference books to find out about the kinds of homes people have in hot desert areas.

  a What are the homes made of?

  b Why are so many houses in desert areas painted white?

  c Why do they often have thick walls and small windows?

  d Collect pictures of different kinds of desert house and make a class display of them.

3 a What kind of clothes do people wear in hot deserts?

  b What colours are the clothes?

  c What materials are the clothes made from?

  d Draw pictures of the kind of clothes people wear in hot deserts.

Tropical rainforests

Some of the wettest places on Earth are **tropical rainforests**, where it is always hot and it rains almost every day. Some tropical rainforests receive more than 4000 millimetres of rain each year. The map on this page shows where the world's tropical rainforests are found.

Why are tropical rainforests important?

Rubber, Brazil nuts, bananas, spices, coffee, cacao (from which some drinks are made) and some medicines come from tropical rainforests. So do some of our most valuable **hardwoods**, including teak, mahogany and rosewood. The plants in the tropical rainforests produce about 40 per cent of the oxygen in the Earth's atmosphere.

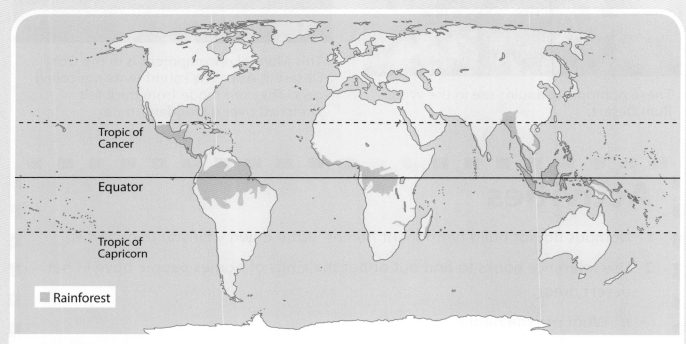

Tropic of Cancer

Equator

Tropic of Capricorn

Rainforest

Tropical rainforests are found in the tropical regions between the Tropic of Cancer and the Topic of Capricorn. Which **continent** has the largest area of tropical rainforest?

Plants and animals in tropical rainforests

The trees in a rainforest are very large and close together. The trees are **evergreen** and, because of the very warm, wet weather, they are able to grow all the year round. There are always flowers, fruits or nuts on the trees for animals to eat, so there are many more kinds of animal in a rainforest than in any other environment. Only about 8 per cent of the world's land area consists of tropical rainforests, but they are home to almost half of the different kinds of plant and animal in the world – so many that some have not even been given names.

Living in the tropical rainforest

Tropical rainforests are home to many small groups of local people, who are experts at living in the rainforest and getting everything they need from it. Some clear small patches of forest to grow their crops and keep their animals. Others are hunters. These people cause very little damage to the rainforest by the way they live.

Rainforests under threat

People are destroying the tropical rainforests. Thousand of trees are being cut down for their valuable timber. Huge areas are being cleared to make way for mines, factories, roads and large animal farms called ranches, or huge **plantations** where only one crop plant is grown. The people and wildlife that live in the rainforest are losing their homes. Most scientists believe that the destruction of the tropical rainforests is also affecting the world's **climate**. While they are growing, the trees remove carbon dioxide from the air to make their food, and carbon dioxide is one of the so-called greenhouse gases that are making the Earth warmer.

A tropical rainforest is made up of different layers of trees, shrubs and other plants.

Did you know?

The Amazon rainforest in South America is home to 20 per cent of the world's birds, 2.5 million different kinds of insects, and 40 000 different kinds of plants.

Activities

1   Look at the graph on the right. Manaus is a city in the Amazon rainforest in Brazil. How is the climate of Manaus different from the climate where you live?

2   Discuss, with a friend, why so few people live in the tropical rainforests.

3   How does your local area compare with a tropical rainforest area? Think about the weather, plants, animals, landscape and people.

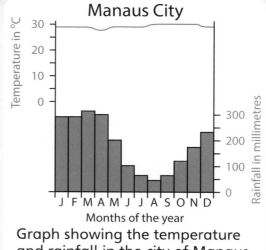

Graph showing the temperature and rainfall in the city of Manaus.

Drought

Some parts of the world have too much rain and others have very little. What happens when there is too little rain? A **drought** is a long period of dry weather, when no rain falls for weeks, months or sometimes even years.

Regular droughts

Many parts of the world have regular droughts every year. These places have a dry season and a wet season. The countries around the Mediterranean Sea, for example, have rain in winter and droughts in summer. People plan for dry periods by storing water and by growing crops that can survive in dry weather. Deserts have droughts all the time, and desert people are experts at surviving in dry conditions.

Unexpected droughts

Unexpected droughts cause the most damage. If it doesn't rain in the wet season, there is no water to store for the dry season.

Droughts cause rivers, lakes, reservoirs and wells to dry up. Plants wither and die. Animals trample the ground down hard as they search for plants to eat and water to drink. The animals die, or their owners have to kill and eat them. The soil becomes dry and dusty and is blown away in strong winds. In the hot sunshine, the dry plants may catch fire. Even when the rains return, crops may not grow well because the soil has been blown away.

With no crops for food, and no animals left to provide meat or milk, country people starve unless they can get food from somewhere else. People who live in towns and cities are affected, too, because when food is in short supply, prices rise and it becomes too expensive for some people to buy the food they need.

Crops, such as these maize plants, wither and die during a drought.

During a long drought, animals die of **starvation** and lack of water.

Drought and famine

A long period of drought can cause a **famine**. A famine is when a large number of people in an area do not have enough to eat. Famine causes people to become weaker and weaker, and they are unable to fight off disease. Many babies, children and adults die of starvation and diseases.

Between July 2011 and June 2012, the worst drought in 60 years affected the whole of East Africa. It caused serious food shortages across Somalia, Djibouti, Ethiopia, Kenya, Sudan, South Sudan and parts of Uganda. As many as 260 000 people died. Many **refugees** from southern Somalia fled to neighbouring Kenya and Ethiopia, where crowded, dirty conditions, and severe food shortages, led to even more deaths.

India has suffered from many droughts and famines. A severe drought struck the Orissa region of India in 1998 when, after several years with little rain, the **monsoon** rains failed to arrive. Crops withered and died and more than half a million people had to leave their homes in search of food elsewhere. Many starved to death.

These people are waiting for food and medicines at a famine relief camp in Somalia.

Did you know?

From 1931 to 1938, a severe drought hit the Southern Great Plains of the United States. To plant crops, the farmers had ploughed up the grasslands that kept the soil in place. As the land dried up, the soil blew away. The area became useless for farming and was named the Dust Bowl.

Activities

1 On a large map of the world, label areas where drought and famine are in the news. Draw a key to your map.

2 Imagine there is a drought where you live. Write down as many ways as you can think of to save water.

3 Some newspapers, television channels and Internet sites show the daily rainfall in large towns and cities across the world.

   a Choose five different cities and write down the daily rainfall in each place, for one week.

   b Write the cities, in order, from wettest to driest.

   c Use an atlas to help you label these places on a map. Where are the wettest cities? Where are the driest ones? Is there a pattern?

Safe water

Is the water that you drink safe? We tend to take water for granted. We can turn on a tap and have water to drink or wash with. We flush the toilet, and water washes away our waste. However, at least two thirds of the people in the world do not have a tap inside their home. They have to get water from a tap, well, waterhole, stream or river some distance away. Their toilet may be a bucket, a hole in the ground, or a river or stream.

How much water do we use?

How much water we use depends on how easy it is to get it. In the poorer countries of the world, many people have to make do with only a few litres a day. This water may have come from a river or waterhole that has been used as a toilet by people and animals.

Water used per person per day	
Africa	30 litres
India	135 litres
Europe	200 litres
United States	575 litres

Fetching water

How do we get our water? In poorer parts of some countries, it is the work of women and children to fetch water for the family to use. For most women, carrying water takes up several hours every day. In the very dry areas south of the Sahara Desert, for example, on average women walk

2–3 hours each day to a waterhole 12 kilometres from their home, and back again. On the return journey they carry the water on their head in a clay pot that weighs about 25 kilograms. In parts of East Africa children have to get up at 3 a.m. and walk 12 kilometres to fetch water for the family.

In country areas of India, a woman may work for 20 hours each day, spending, on average, about 5 hours of this fetching and carrying water.

Water and disease

For many of us, the water that comes from our taps is clean and free from germs. For people in poorer countries, the water they drink, cook or wash in can make them ill and cause diseases such as cholera, typhoid, dysentery and diarrhoea.

Scientists believe that eight out of ten people who are ill in the poorer hot countries have a disease that has been spread by the water they drink or use.

Every day about 70 000 children die from diseases spread by drinking or using water that has been **polluted** by sewage.

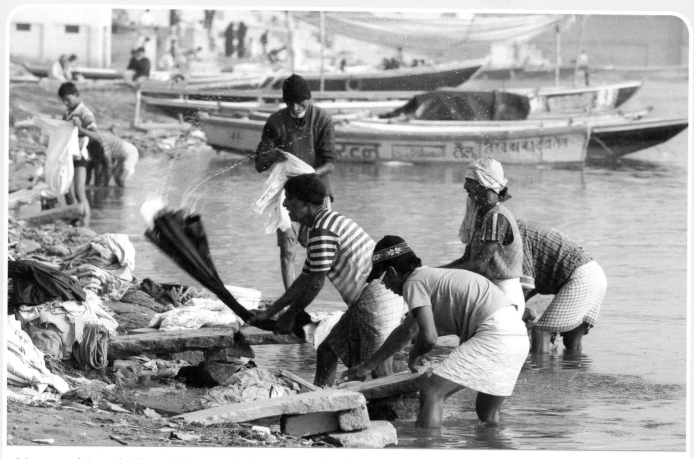

Men washing clothes, dishes and themselves in the **polluted** waters of the Ganges River in India.

Did you know?

One in three people in the world do not have the use of a toilet, and one in eight people do not have clean water to drink.

Activities

A terrible disease called cholera struck parts of London, England, in 1854. Hundreds of people died. A local doctor, John Snow, was able to prove it was because of the dirty water people drank. Find out more about cholera from reference books or the Internet. Find the names of two countries where people still suffer from cholera.

Who owns water?

Although the water you use originally fell from the sky as rain, you have to pay for it, unless you get your water from a well in your own garden. Most people who live in towns and cities must pay the local water company or the government for the water they use in their homes. Many homes have a meter that measures how much water has been used.

The cost of clean water

Why do we have to pay for water? It costs a lot of money to build desalination plants or dams and reservoirs, waterworks, water towers and all the pipes that carry water to our homes. It also costs a lot to clean and make the water that comes from our taps safe to drink. Taking the salt out of sea water is very expensive. When we have used the water, we have to pay towards the cost of the pipes and sewage works that clean the water and make it safe to put into a river or the sea.

Unsafe water

In the poorer countries of the world, there are few water companies. Even if there were more, many people would not have enough money to pay for clean water. Many people have to take their water from polluted wells, streams and waterholes. In these places, too, few homes or villages have a safe toilet. Often, people have to use an open pit or a bucket that may seep or overflow into the local well, river or stream. This makes the water unsafe to drink.

These children in Allahabad, India are lucky. They can wash themselves under a tap.

Did you know?

There is a lot of water deep underneath the Sahara and Arabian Deserts, which has been there for thousands of years. It is called 'fossil water'. Some water from rainfall is added to this 'fossil water' each year, but it is being used up, mainly to **irrigate** crops, faster than new water is added.

Charities and water

Many **charities**, such as Oxfam, UNICEF and Water Aid, are working to try to provide clean water and safe toilets for the people in poorer countries. They do this by drilling wells deep into the ground, in places where the underground water is clean and safe to use. Unlike the wells built by local people, which are often shallow and dug by hand, these new wells are lined with piping and covered so that the water does not become polluted.

In countries where there is some rain, small dams and reservoirs are being built to collect rainwater. The charities also build gutters and special cement tanks to collect the rainwater that falls on roofs. In a few places, solar panels, which produce electricity from sunlight, are used to drive pumps that bring up water from deep underground.

The water coming from this tap in a village in India was collected from rain and stored in the large tank behind the woman.

This water pump in The Gambia is powered by solar panels.

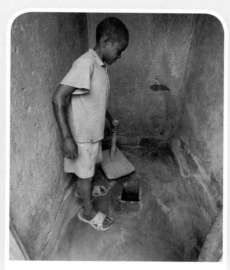

Charities also build pit **latrines** in country areas. A pit latrine is a kind of toilet used where there are no **sewers** or running water.

Activities

1. Design a poster to help to raise money for a water project in a country where there is a shortage of clean, safe water.

2. Look at a water bill for your home. How is the amount you have to pay worked out? What kinds of things does the water company charge for?

3. You can work out how much water a dripping tap wastes. Turn on the cold tap until it is just dripping. Put a bucket under the tap and measure how much water drips into the bucket in an hour. How much water would drip away in, **a** a day **b** a week and, **c** a year?

Forms of transport

In the past, most people walked or travelled by horse or some other animal when they went from place to place. In some parts of the world these are still the only methods of travel for many people.

Different forms of transport

In much of the world the most common ways of transporting people or goods are bicycles, buses, vans, trucks, trains, aeroplanes and ships.

The different forms of transport follow fixed routes to get from one place to another. Motor vehicles travel along roads, trains run on railway lines, aircraft follow flight paths, and ships keep to shipping lanes in the oceans and seas.

Why is transport important?

There is now hardly anywhere in the world we cannot reach in 24 hours. A good transport system is important because it allows:

- people to move from place to place for work, school and holidays

- emergency services, such as ambulances, fire engines and police to get quickly to places where they are needed

- letters and parcels to be sent and received

- raw materials to be carried to farms and factories

- food and other goods to be delivered from farms and factories to shops.

Where there are roads, motor vehicles can take people and goods right to the place they need to go. However, motor vehicles are noisy, they pollute the air and can be dangerous, particularly to **pedestrians** and cyclists.

Trains link big towns and cities and can carry lots of **passengers** or heavy cargo over long distances. However, they are noisy and pollute the air. There are many places that trains cannot reach because there are no railway lines.

Bicycles are inexpensive, quiet and they do not **pollute** the air. However, they are not good for long journeys and they are not very comfortable in bad **weather**.

Ships and boats can cross oceans and seas carrying lots of passengers and large quantities of cargo. However, they are slow and can travel only from one **port** to another.

Land and buildings

Some transport needs special buildings. Bicycles and motor vehicles need roads or motorways. Trains need railway lines and railway stations. Ships need harbours and docks, and airplanes need airports with road and rail links to them.

Airplanes are the fastest form of transport. They can carry large numbers of passengers and travel across land, oceans and seas. However, they are noisy and pollute the air. They cannot carry very large or very heavy cargo. There are not many airports, so people may have to travel a long way by road or rail to an airport.

Activities

1. a Write one advantage of motor vehicles over trains for passenger travel.

   b Write two disadvantages of rail travel.

   c Is train travel best for short, medium, or long journeys?

   d Write two advantages of trains over trucks for transporting goods.

2. a Which is the nearest airport to your home or school?

   b Use maps and local information or the Internet to find out where the main roads are that people use to get to and from the airport.

   c Do the roads pass through towns and villages, or do they go around them?

Road traffic

Roads are special surfaces for cars, trucks, buses and other motor vehicles to travel on. Roads vary in size and importance.

What were early roads like?

The first roads were bumpy, twisting, narrow tracks used by people and their herds of animals. They were muddy in wet weather and hard and dusty in dry weather. The first proper roads were built about 6000 years ago. They consisted of stone paved streets at Ur, in modern-day Iraq, and timber roads preserved in a swamp in Glastonbury, England.

About 2000 years ago the Romans built straight roads in Europe and North Africa. These roads were needed so that soldiers could travel quickly from one part of the Roman Empire to another. Roman roads were paved with stones and had a sloping surface to let rainwater drain away.

A Roman road crossing the Gredos Mountains in Spain.

What are roads like today?

In many parts of the world, country roads still follow the routes of the first tracks. They twist and turn around farms and homes and other obstacles such as hills and forests. In towns and cities some roads are narrow because they were built before the days of motor vehicles. However, most modern roads are built so that they travel in almost straight lines across open country. There are no sharp bends, steep hills, crossroads, roundabouts or traffic lights. Busy built-up areas are avoided, or passed through by means of tunnels or raised-up sections of the road. There are often two or more lanes in each direction and the road is made safer by separating the traffic going in opposite directions with a central barrier. Motorists can only join or leave the road where it meets another large road. Where that happens, ramps connect the two roads so that motorists can join the connecting road without disturbing the flow of traffic.

To make it safer to drive at night, there are reflective signs, overhead lighting and reflective markers, often called 'catseyes', to separate the different lanes of traffic. Electronic signs give warnings of traffic problems ahead.

This busy road junction is in Saudi Arabia.

Did you know?

In the United States motorways are called 'highways' or 'freeways'. In France they are called *autoroutes*, in Italy they are called *autostrada*, and in Germany they are called *autobahns*. Other names for motorways are 'national routes' and 'expressways'.

Roads and development

Good roads are important for the businesses of an area. Factories and shops need a good road **network** to move the things they make or sell. If an area has good roads, new factories, offices and shops may be built there, which will create jobs for people.

Most factories and many supermarkets and other large stores are now built on the outskirts of towns and cities. There is less traffic in these areas, and it is easier to deliver goods and materials to and from factories and shops.

Activities

1 Look at a map of your local area.

   a What types of road are there?

   b Which towns and cities do they pass through?

   c How long is each road?

   d Record your results in a spreadsheet.

2 Draw a map of the area where you live. Include all the different types of road. Label the roads with their correct names. Draw a key for your map.

Traffic problems

In most countries today, people use cars more than any other kind of transport. Trucks and vans are the main method of moving goods from place to place. However, an increasing number of motor vehicles on the roads brings traffic problems.

Traffic jams

The narrow streets of old towns and cities were not built for large amounts of traffic or for large trucks. As the number of motor vehicles has increased, many towns and cities have traffic jams. Traffic jams make motor vehicles travel very slowly.

There are even traffic jams in country areas, where the roads are narrow and not built for today's modern trucks or cars.

Rush-hour traffic in Alexandria, Egypt.

Noise and vibration

Motor vehicles of all kinds cause noise pollution. The noise can be very bad in crowded towns and cities, and for those people who live near busy fast roads and motorway. The vibration caused by large trucks can even damage buildings.

Did you know?

In 2012 in the centre of London, England, the average speed of the traffic was 15 kilometres an hour. This is slower than it was in 1900 when horses were the main method of transport.

Air pollution

Motor vehicles also pollute the air with fumes from their exhaust pipes. This is what these fumes contain:

- **Soot**: These small particles dirty buildings and damage people's lungs if they are breathed in.

- **Carbon monoxide**: This gas makes your lungs unable to take oxygen into your body. It also helps to cause a dangerous mixture of smoke and fog, called smog.

- **Sulphur dioxide and nitrogen oxides**: These gases dissolve in water droplets in the air and help to cause acid rain. They add to **global warming** by trapping the Sun's heat. These gases also cause breathing problems.

- **Hydrocarbons**: These gases may cause cancer and breathing problems. They also help to cause global warming.

- **Carbon dioxide**: This is a greenhouse gas. This means it traps heat from the Sun, helping to cause global warming.

Why is this police officer wearing a mask?

Accidents and asthma

It is believed that 300 million people worldwide now suffer from asthma or other breathing difficulties, which many scientists think are caused, or made worse, by air pollution. In addition to the noise and air pollution caused by motor vehicles, many people are killed or injured in road accidents.

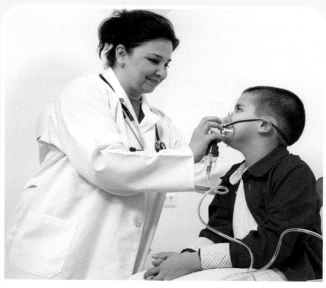

Many doctors and scientists believe that the increase in the number of people with **asthma** around the world is caused by air pollution.

Activities

1. **a** Many large cities around the world have serious problems caused by road traffic. Choose one of the following cities and use reference books or the Internet to find out what traffic problems the city has: Cairo, Dubai, Athens, Bangkok, Tokyo, Mexico City, Los Angeles and Jakarta.

   b Find out what is being done to solve the traffic problems. Write a paragraph on what you found out.

2. Traffic accidents are a major problem caused by motor vehicles. Find out how many serious traffic accidents resulting in death or injury occurred last year in your local area.

Solving traffic problems

People in many countries have become very worried about the effects of traffic in towns and cities. They are concerned about:

- the harmful effects on human health

- the damage to the environment caused by the noise, vibration, acid rain and global warming

- the waste of time and of **non-renewable fuels** such as petrol and oil in traffic jams

- the use of valuable space in city centres for parking cars and trucks.

In the past, governments believed that building bigger, straighter roads was a way to solve traffic problems. However, even new roads soon filled with more and more motor vehicles.

Better forms of transport

One way to solve traffic problems is to get people to use cars and trucks less. Walking and cycling are good for short journeys. They give you healthy exercise and cause no pollution at all. Buses, trams and trains can carry many people at once and cause a lot less pollution per person than cars.

Using fewer cars

Various ways are being tried to persuade people to stop using cars, or to use them less. These include:

- increasing the tax on petrol and diesel oil to make it more expensive to use a car

- making parking in towns and cities more difficult and more expensive except for those people who actually live in that town or city

- not letting cars and trucks into some towns and cities at busy times of the day

- closing some shopping streets to motor vehicles, so that they can only be used by **pedestrians**

- park-and-ride schemes that allow people to park in car parks on the outskirts of towns and cities and then travel into the centre by bus

- traffic-calming schemes, such as ramps and chicanes, to force motor vehicles to travel more slowly through streets where people live, shop or go to school

- trying to encourage more companies to send their goods by train instead of by trucks.

Motor vehicles are not allowed on this city street in Dubai.

Subway and Metro systems

A few cities, including London, New York and Paris, have a large network of underground or subway trains. Such a network is very expensive and difficult to build, but it reduces the number of people travelling by road. Other cities such as Cairo, Tehran, Kolkata, Seoul, Dubai, Brussels, Brasilia, Moscow and Beijing have built special electrically driven light railways or Metro systems to carry passengers from one part of the city to another quickly, safely and with little noise and pollution.

The Metro system in Seoul, South Korea, is the world's longest subway system and has the most stations.

Shanghai in China has a monorail Metro system, which carries passengers between the airport and the city centre. It is capable of travelling at speeds of up to 500 kilometres an hour.

Activities

1 Imagine that you are a shopkeeper in a street from which motor vehicles have been banned to reduce traffic problems.

   a Give one reason why you might be pleased about this.

   b Give one reason why you might not be pleased.

2 One way to reduce the number of cars in large towns and cities is to make drivers pay a **toll fee** every time they enter the town or city. Another is to start a park-and-ride bus scheme.

   Which plan do you think would be best and why?

3 In the rush hour in many towns and cities, each car often contains only one person.

   a Work with a friend and discuss what could be done to change this in order to reduce traffic problems.

   b Write down your conclusions.

Do we want a new road?

Many cities around the world have a **ring road** to take the traffic out of the city centre. They include Beijing, Cairo, Johannesburg, Hong Kong, Kuala Lumpur, Lahore, Singapore and Berlin. A **bypass** road is sometimes built to take all the through traffic away from a town or city. Such a road makes people's journeys quicker and safer, while life in the town or city is a little safer and less polluted.

Once such a ring road or bypass is complete, new factories and offices are often built near it, bringing new jobs to the area. Not everyone welcomes a new road though.

Part of the ring road in Johannesburg, South Africa. Notice how new factories have been built by the road.

New roads and traffic problems

New roads can bring more cars and trucks, causing more noise and pollution, because better roads make journeys easier. When a motorway was built all the way round the outskirts of London, England, the engineers thought it would carry 80 000 vehicles a day. Within a few months it was carrying more than 160 000 vehicles a day, causing traffic jams and even more noise and pollution. The motorway has had to be widened in places to take all the extra traffic.

Did you know?

The world's longest ring road is the Berliner Ring, which goes all the way around the German capital city of Berlin. It is 196 kilometres long.

New roads and land use

New roads take up a large area of land. It may be necessary to knock down lots of homes that are in the way of the new road. Other homes may have the new road close to them, causing noise and pollution for the owners.

A new ring road or bypass may use up valuable farmland, or even cut a farm in half so that it is difficult for the farmer to use the land. The new road may also cut across beautiful countryside or destroy forests, marshes, grassland and other areas where rare or unusual plants and animals live.

New roads take up valuable farmland and destroy countryside and wildlife.

New roads and local businesses

Not everyone in a bypassed town or village will welcome the new road. Shopkeepers and the owners of garages, cafés, restaurants and other businesses may lose trade because there is less through traffic. So, although it might reduce the traffic in a town or village, a bypass is extremely expensive and not everyone would like it. What do you think?

Activities

1 Look at a large-scale map or road map of your area. Write a list of all the villages, towns and cities that have been bypassed.

2 Imagine there is a proposal to build a bypass around the area where you live.

   a Work with a group of friends and discuss whether, and where, the bypass should be built.

   b Play the parts of the following local people in your discussion: a farmer, an elderly person, a traffic planner, a school head teacher, a truck driver, a birdwatcher and the owner of a house on the **route** of the proposed bypass.

3 Imagine you are a local planning officer. You receive a proposal to build a large new shopping centre and a cinema complex next to the local bypass. Write a list of arguments for and against allowing the shopping centre and cinema complex to be built.

Coastal features

What is a **coast**? A coast is a place where the land meets the ocean and sea. A map of the world is really a map of the coasts of the various continents and countries. The world has about 312 000 kilometres of coastline. Have you visited any places on the coast? Find them in an atlas or on the Internet.

The seashore

The **seashore** is the land along the coast. Some seashores have a strip of sand, shingle, pebbles or mud at the edge of the sea. This is called a beach. Sometimes there are cliffs or sand dunes behind the beach.

The **wind** and **waves** have shaped this coastline in Australia into a series of cliffs, caves, bays and rocky towers called **stacks**.

Changing coastlines

The shape of the coast changes all the time because of the action of the waves and wind. Where the rocks are soft, the waves wear away the land, moving the coast inland. Where the rocks are hard, there may be little change. In some places, the waves form new land along the seashore by bringing sand and tiny pieces of rock from other parts of the coast.

Some of the largest inlets on the coast are **estuaries**, where rivers flow into the sea. Tiny pieces of mud and sand swept along by the river gradually settle on the bottom where the river's **current** is slowed when it meets the sea. Over many years, the particles form large areas called mudflats. Eventually these mudflats may dry out and form new land.

An estuary is the mouth of a river, where it flows into the sea. This estuary is in Wales.

Why are coasts important?

We use sheltered parts of the coast as **ports** for ships and fishing boats. A lot of our food, including fish and shellfish, comes from places on the coast. Salt, oil and natural gas are found in coastal areas. Many people go to the coast for their holidays or to enjoy sports and leisure activities.

How people have harmed the coast

People damage the coast in many ways. They leave litter and they damage sand dunes by walking or driving over them. When oil, **sewage** and harmful chemicals from factories, towns and cities are pumped into the sea, this **pollutes** the beaches and seawater. It harms the plants and animals living along the coast and makes the seawater dangerous to bathe in.

Many people like to go to the coast for their holidays. It is important for their health and safety that the beach and sea are clean and unpolluted.

Did you know?

The world's longest estuary is that of the River Ob in Russia. It is 885 kilometres long and up to 80 kilometres wide.

Activities

1 a Collect pictures of different kinds of coastline.

  b Make a class display or a scrapbook with your pictures.

  c Write a sentence or two to describe each picture.

2 a Make an illustrated dictionary of the following words to do with coasts: beach, cliff, wave, sand dune, pebble, estuary, mudflat, port.

  b Add new words to your dictionary as you learn them.

3 a Use a map in an atlas or on the Internet to choose a part of the coast you would like to visit.

  b Write down the reasons for your choice.

The movements of the ocean

The oceans and seas are never still. The water moves all the time because of **tides**, waves and currents.

Tides

The level of the ocean rises twice each day and water covers the shore. We say that the tide is 'in', or that it is 'high tide'. Twice each day, the level of the water falls. The seashore is uncovered and we say the tide has 'gone out', or it is 'low tide'. High tide brings deep water to **harbours** and ports so that large ships can sail in and out. Sometimes a very high tide can cause flooding.

Tides are caused mainly by the pull of the Moon's gravity as it circles the Earth. This pull causes the ocean's water to pile up in a bulge on the side of the Earth facing the Moon. Another bulge of water forms on the opposite side of the Earth. As the Earth spins, places get high and low tides as they move in and out of these bulges of water.

Spring tides

Spring tides occur twice each month when the Sun and Moon are both on the same side of the Earth. Their combined pull makes the ocean's water pile up higher than usual.

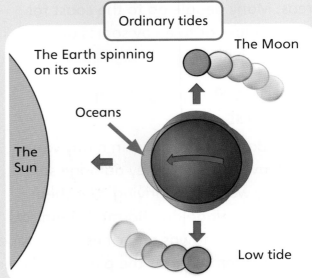

Ordinary tides

The Earth spinning on its axis

The Moon

Oceans

The Sun

Low tide

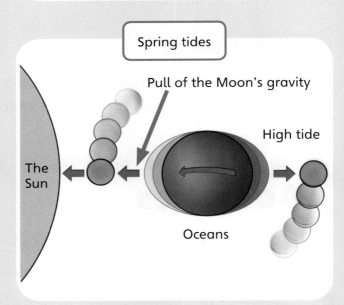

Spring tides

Pull of the Moon's gravity

High tide

The Sun

Oceans

How the tides are formed.

Waves

Waves are big ripples of water caused by the wind. They do not move the seawater from place to place like currents or tides. Out at sea, waves look as if they are moving forwards, but really the water in each wave stays in almost the same place, moving round in circles. The longer and stronger the wind blows, the bigger the waves will be. Near the shore, some of the water at the bottom of the wave rubs against the seabed. This slows the bottom of the wave down and its top curves over and breaks.

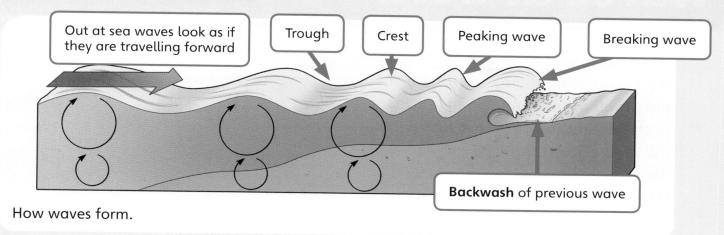

Out at sea waves look as if they are travelling forward

Trough

Crest

Peaking wave

Breaking wave

Backwash of previous wave

How waves form.

Currents

Currents are like giant rivers of water moving slowly through the oceans. Most currents are caused by winds that blow over the surface of the ocean in the same direction all the time. However, cold water near the polar regions sinks to the bottom of the ocean, making a current deep down. Warm water moves in to take its place, forming another current.

Activities

I Read these statements about the Mediterranean Sea and answer the questions that follow.

- Over most of the Mediterranean Sea the rise and fall of the tides between high and low water is less than 30 centimetres.

- There are no strong currents.

- **Pollution** has become a serious problem.

   a Find out the reasons for the three statements above.

   b Is there a connection between these three statements? If so, what is it?

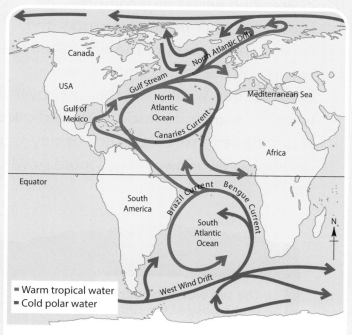

The main currents in the Atlantic Ocean.

Waves at work

Waves carry with them some of the energy and power of the wind. When large waves reach the seashore, they can hurl pebbles and rocks against the cliffs and sea walls. This erodes, or wears away, the cliffs. Waves can also damage buildings and other structures along the seashore.

How do waves erode cliffs?

Cliffs along rocky coasts often have strange shapes. These shapes were formed by the action of the waves. As the waves throw pebbles and lumps of rock at the base of a cliff, they gradually erode it. Breaking waves also **compress**, or squeeze, the air trapped in cracks and crevices in the rocks of the cliff. When the waves pull back, the compressed air

Waves gradually wear away, or erode, the bottom of a cliff.

explodes out, weakening the rock. Over time, pieces break off the cliff. Seawater weakens and erodes rocks such as chalk and limestone by slowly **dissolving** them. A cliff made of soft rock, such as clay, may erode quite quickly. Harder rocks, such as granite, wear away much more slowly.

1 The sea erodes a hollow at the base of the cliff so that the upper parts of it overhang. This weakens the cliff and large sections of it may crumble into the sea.

2 The whole process then starts all over again, and the cliff face moves further and further back.

3 A flat area, called a wave-cut platform, is left where the cliff once stood.

How a cliff is worn away to form a wave-cut platform.

Structures formed by waves

Waves form different structures when they erode cliffs. Study the diagram below. It shows how bays, headlands, sea caves, blowholes, arches and stacks are formed.

1 The sections of a cliff that are made of soft rock erode quickly, forming a bay or inlet.

2 The sections of a cliff that are made of hard rocks take longer to erode and are left jutting out as headlands.

6 When the roof of the arch collapses, it leaves a column of rock called a stack. In time the stack will also be worn away by the waves.

5 Caves on opposite sides of a headland may meet to form an arch.

3 In places where the rock is weak, waves can erode a deep hollow into the cliff, forming a cave.

4 Waves may erode a hole in the roof of the cave, called a blowhole.

Structures formed by waves.

Did you know?

The world's biggest bay is Hudson Bay, Canada. Its shoreline is 12 268 kilometres long.

Waves have formed this headland, arch and rock stacks on the coast of Portugal.

Activities

1 Collect pictures showing bays, headlands, arches, stacks, caves and blowholes formed by waves and make a class display or scrapbook. Include a sentence or two to describe each picture.

2 Some rocks are harder than others. Think of a method to compare the hardness of different types of rock. Try out your method to see if it works. Share your results with the class.

Beaches

What is a beach? Beaches are the gently sloping areas of sand, pebbles or mud at the edge of an ocean, sea or lake.

Where the coast is sheltered, the beach is often made of sand. At low tide there may be seaweed-covered rocks and rock pools. On open, wind-swept coasts, the beach is usually made of pebbles or rocks, because most of the sand has been washed away.

Where does beach sand come from?

Most sand is made up of tiny particles, or grains, of rock. When the waves erode cliffs, lumps of rock break off and crash down to the beach below. These large pieces of rock are rolled backwards and forwards by the waves and tides, along with smaller pebbles. Pieces break off the boulders until they are worn down to pebbles and then, eventually, into tiny grains of sand.

The **boulders** that have fallen from this cliff will eventually be broken down by wind and waves to form grains of sand.

Bricks, pieces of concrete and glass on a beach are slowly worn smooth by the action of the waves, so that they eventually form tiny grains of sand.

These seashells will eventually be worn away to become part of the sand on the beach.

Rivers bring millions of tonnes of sand and mud to the oceans and seas every year. This material has been eroded from mountains, hills and rocks inland. This sand and mud may be swept along the coast by tides and currents and eventually be **deposited** on a beach.

In the **tropics**, many beaches are made almost entirely of tiny fragments of seashells or **coral** washed up from nearby reefs.

Sorting beach materials

On many beaches at low tide, you will find that there are large pieces of rock and pebbles at the top of the beach and smaller pebbles and sand down near the water. How was this material sorted?

When a wave breaks on the shore, it has a lot of energy and the water rushes up the beach, taking rocks, pebbles and sand with it. The water drains back down the beach in what is called the **backwash**. The backwash is weak and cannot carry the rocks and larger pebbles far, so these are left further up the beach. Smaller pebbles are dropped next, and finally tiny particles of sand down near the water's edge.

At the back of some open, wind-swept beaches are very large rounded boulders. These were thrown there during violent storms. That is why these beaches are called storm beaches. This storm beach is in western Ireland, by the Atlantic Ocean.

Did you know?

Although it may look bare, a sandy beach may be hiding 2000 shellfish and other small animals in every square metre of sand.

Activities

1. Boulders break down into pebbles and then into grains of sand. You can show this with the following experiment:

   Step 1: Place a few small pieces of blackboard chalk in a plastic jar half full of water.

   Step 2: Screw the lid on the jar and shake it vigorously.

   Step 3: What happens to the pieces of chalk?

   Step 4: What is left in the bottom of the jar?

2. Beach sand is made of tiny pieces of rock, shells or coral.

   a Write down as many uses of beach sand as you think of.

   b Use reference books and the Internet to find out the uses of sand.

3. a Use holiday brochures, the Internet and holiday guides to find out where the best sandy beaches are in the world.

   b Write lists of the sandy beaches arranged by continent or country.

Types of coastal settlement

Use an atlas or the Internet to find how many of the world's largest cities are on the **coast**. As well as cities, there are also large **ports**, industrial towns, villages and holiday **resorts** on the coast. By building and developing these settlements, people have changed the environment of parts of the coast.

Why do settlements develop on the coast?

- **Natural harbours**: Parts of the coast are sheltered from the open sea and have natural deep-water harbours. These are the best places to launch a boat and so villages, and later ports, have developed in places like this. Portsmouth in southern England, Sydney in Australia, New York in the United States, Kingston in Jamaica and Vancouver in Canada all developed around natural harbours.

- **Estuaries**: An estuary is where a river widens out as it reaches the sea. Estuaries are excellent sites for ports because they are sheltered from the main force of the waves and tides. Some of the biggest ports in the world, including Dublin in Ireland and Buenos Aires in Argentina, developed in the shelter of river estuaries.

- **River crossings**: Villages also grew into towns and cities near the mouths of rivers, where they could be crossed by a bridge or ford. In these places, where land and sea routes joined, large cities often developed, including London, Lisbon, Montreal and Alexandria.

- **Important resources and industries**: Other coastal towns and cities grew because there were important **resources** or **industries** nearby. San Francisco in the United States grew into a city and port because gold was found nearby. Tromso in Norway started life as a port based on fishing and the capture of whales. Newcastle in Australia grew into a major port city because large supplies of coal were found nearby.

The city of Perth in Australia, and its port Freemantle, developed rapidly after gold was discovered nearby in 1893.

- **Holiday resorts**: Beautiful places along the coasts, especially those with wide sandy beaches, sometimes develop into holiday resorts. They include some of the world's best-known resorts, such as Miami Beach in Florida, Sharm el Sheikh in Egypt, Dubai in the United Arab Emirates, Acapulco in Mexico and Benidorm in Spain. Which coastal holiday resorts do you know?

Why do you think this holiday resort in Brazil grew? Do holiday resorts damage the environment?

How do ports affect the environment?

With the development of a port comes the growth of oil refineries and industries that use the port to bring in their **raw materials** or to **export** their products. However, as a port develops, the habitats of plants and animals are destroyed by the building work and pollution. This pollution can make nearby beaches unusable for tourists and holidaymakers.

Tromso in Norway began life as a port based on fishing and the capture of whales.

Activities

1 a Use an atlas or the Internet to find the estuaries of the world's major rivers, such as the Amazon, Nile, Ganges, Volga, Yangtze, Niger and Congo.

   b Name the large towns or cities on each of these estuaries.

   c Which rivers flow into these estuaries?

   d Can you find any large bridges or tunnels crossing these estuaries?

2 Some newspapers record the number of hours of sunshine each day for holiday resorts. Choose five resorts and write down the number of hours of sunshine for each resort for a month. Draw a bar chart of your results. Which is the sunniest seaside resort in your sample?

A coastal holiday resort

Benidorm, on Spain's south coast, is Europe's biggest holiday resort. Can you find Benidorm on the map below? Benidorm has a warm, sunny climate all the year round, thanks to the mountains which shelter the town. The average temperature in winter is 15°C, and 26°C in summer.

Benidorm is a seaside resort on the south-east coast of Spain.

A fishing village

Before the 1950s, Benidorm was a small fishing village. Fishermen in Benidorm specialised in catching tuna, but in 1952 the biggest tuna fishery there closed down because the number of fish being caught fell.

Members of the town council looked for ways to encourage tourists to visit Benidorm. They wanted people to enjoy Benidorm's pleasant climate and sandy beaches that stretch for 6 kilometres along the coast. Until 1959, Benidorm only had

four hotels, but soon many new hotels and apartment blocks were built. Benidorm was one of the first holiday resorts to have 'package holidays'. This is where flights and hotel or apartment accommodation are sold together, and special flights, called charter flights, carry the tourists to their destination.

As Benidorm had excellent travel connections, by land, sea and air, many holidaymakers travelled there from all over Europe. The nearest airport, at Alicante, was only two hours flight away from most European capital cities, such as London, Paris, Brussels, Amsterdam, and Berlin.

Playa Levante, one of Benidorm's sandy beaches, is backed by huge hotels.

Did you know?

With more than 8000 kilometres of sandy beaches, Spain is one of the most popular tourist destinations in Europe.

Today, Benidorm is the most popular holiday resort in Europe, with 5 million tourists visiting every year.

Tourist attractions at Benidorm

The things that attract tourists and holidaymakers to Benidorm include:

- the warm climate, warm seawater and sandy beaches

- leisure activities such as sailing, walking, horse riding, cycling and boat trips

- four theme parks – one, called Terra Natura, shows visitors the wonders of the animal kingdom; and Terra Mitica, the largest theme park in Europe, has rides, and shows that include acrobats, stuntmen and circus acts.

The roller coaster at the Terra Mitica theme park is 1252 metres long and 36 metres high.

Activities

1 a Find Benidorm on a map or in an atlas. How far away is it from your home? In which direction does it lie from your home?

   b What is the nearest seaside resort to your home? How far away is it and in which direction?

2 a Use the Internet to find out how the temperature in Benidorm varies throughout the year.

   b Draw a line graph of your results.

   c Which is the warmest month?

   d Draw a line graph showing the number of hours of sunshine in Benidorm each month.

   e Which is the sunniest month?

3 Describe a holiday you have taken at the coast. What was the coast like? What was there to do? If you have never been to the coast, choose a seaside resort and imagine what a visit would be like.

The coastal city of Aqaba

Aqaba is a city on the coast of Jordan at the north-eastern tip of the Red Sea. It is Jordan's only coastal city and its only port. Aqaba first developed about 4000 years ago at the meeting point of two major trade routes that at the time linked Africa and Asia. Today, Aqaba is famous as a holiday resort and port.

Aqaba holiday resort

Aqaba is one of the main tourist resorts of Jordan. Its tourist attrations include:

- a pleasant desert climate, with warm winters and hot, dry summers
- sandy beaches and warm seawater, which are enjoyed by bathers and windsurfers
- the rich sea life which attracts scuba divers and snorkelers
- trips in glass-bottomed boats to see the corals and fish life
- many hotels, holiday villas, a golf course and marinas for pleasure boats.

Over 200 different kinds of **coral** and about 1000 different kinds of fish live in the **sea** around Aqaba. For people who do not dive or snorkel, there are trips in glass-bottomed boats to see the coral and sea life.

One of the marinas at Aqaba.

The port of Aqaba

Aqaba is a busy **container** port. Most of Jordan's export goods leave the country from this port. It is also a busy ferry port, with ferries carrying passengers, cars and trucks between Aqaba and ports in Egypt. There are many factories around the port. Some make heavy machinery, such as motor vehicles. Others handle chemicals such as phosphate, which is used in industry and as a fertiliser. Phosphate is one of

Jordan's most important exports. Cruise ships regularly use the port of Aqaba.

Transport links

Aqaba is connected to the rest of Jordan by two major highways, the Desert Highway and the King's Highway. There are many bus services between Aqaba and Amman and the other large cities in Jordan. Aqaba also has an airport, the Aqaba King Hussein International Airport, with flights to Amman and several airports in Egypt and Europe.

A railway is being built that will connect Aqaba with all of Jordan's main cities and with several neighbouring countries.

Cruise ships regularly stop at Aqaba.

Did you know?

The Red Sea is one of the warmest, saltiest and reddest seas in the world. It gets its name because it contains algae that are reddish-brown in colour.

Activities

1 a Find Aqaba on a map in an atlas or on the Internet.

   b What is the distance from your home to Aqaba?

   c In which direction does Aqaba lie?

   d How would you get to Aqaba from your home or school?

2 a What are some of the buildings found only in seaside resorts? Think about buildings that protect the coast, buildings used by people who help keep you safe at the coast, and buildings used to entertain people using the coast or beaches.

   b Where are these special seaside buildings likely to be?

3 a Ask your friends which holiday resorts they have visited.

   b Draw a bar chart of your results.

   c Which resorts have been visited most often?

   d Are these the resorts that are nearest to your home area?

The container port of Jebel Ali

The port of Jebel Ali, near the city of Dubai in the United Arab Emirates, is the world's largest man-made harbour and the biggest port in the Middle East. It was built in the late 1970s when the nearby Port Rashid could no longer handle the largest cargo ships. Jebel Ali is the largest container port between Rotterdam and Singapore. It is close to the main east–west trade routes, so it can act as a link between the countries of Europe and the Far East.

Jebel Ali Port

The port of Jebel Ali is not located in a sheltered inlet on the coast or in the mouth of a river. It was built by reclaiming land from the sea. The new land was formed by dredging up sand and other materials from the seabed. The port was then sheltered from the wind and waves by long breakwaters built of rock and concrete.

Large ships, such as bulk carriers, supertankers and container ships, all need deep water and special equipment for handling their cargo. Jebel Ali Port handles large amounts of bulk cargoes, such as oil, gas and chemicals. It is one of the ten busiest container ports in the world. It handles over 6500 ships a year and in 2013 it handled 13.6 million tonnes of cargo.

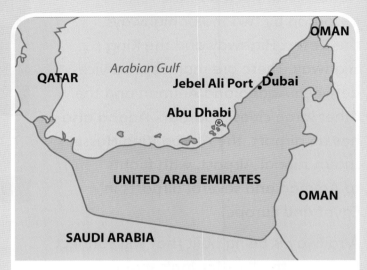

Jebel Ali is a port in the United Arab Emirates.

Jebel Ali Port in Dubai handles over 6500 ships every year.

Jebel Ali harbour is a good cargo port because it:

- is very deep – deep enough for the largest ships in the world

- has berths for 22 large cargo ships at any one time, including 10 of the largest container ships afloat

- has 78 large cranes that allow it to handle up to 19 million containers a year

- has a separate section that deals with roll-on/roll-off ships

- can handle 500 000 cars in a year

- has close links to the cargo area of Dubai Airport, so that small urgent cargoes can be moved from ship to aircraft in just four hours, faster than any other port in the world.

One of the world's largest container ships, the *Marie Maersk*, is 398 metres long and 58 metres wide, and can carry more than 18 000 containers.

Roll-on roll-off ships have large doors at the bow (front) and stern (back), allowing cars and trucks to be driven straight on or off of the ship to load and offload cargo.

Activities

1. a Look at a map showing Dubai and Jebel Ali Port.

   b Why is Jebel Ali in a good position for a port?

   c What other ports share the same coastline along the Arabian Gulf as Jebel Ali Port?

2. Look at a map or atlas and find Dubai. How far away is it from your home and in which direction?

3. Find out what work is carried out at a port by:

   a coastguards

   b immigration officers

   c customs officers.

Singapore, gateway to the East

Singapore is a small island state, made up of over 60 islands. Most of the smaller islands are uninhabited. Can you find Singapore on the map below?

Singapore is an island state off the coast of Malaysia.

Singapore City

Singapore City is the capital of Singapore. It is on Singapore Island, the largest of the Singapore islands. More than 90 per cent of the population of Singapore live in Singapore City. It is a crowded city with many tall apartment blocks, and high-rise offices, banks and hotels.

This is Singapore's main shopping street.

Climate and wealth

Singapore lies just north of the Equator, and its climate is hot and humid with frequent heavy rains. Singapore is so small and so crowded that it cannot produce enough food for all its people. Most of its food has to be imported from other countries. Singapore's water supplies are piped in from Malaysia. As there is a shortage of land, marshy ground is being drained to provide more space for homes and factories. However, in spite of its small size, Singapore is one of the richest countries in South-East Asia. Its wealth comes from shipping, international trade, electronics manufacture and banking.

The Port of Singapore

Singapore is the world's second-busiest port, after Shanghai in China. It is a busy port because it:

- has a sheltered, deep-water harbour
- is positioned between the Indian Ocean and the South China Sea
- has berths for ships, over 200 cranes, and warehouses and storage areas
- can handle cruise ships and the world's largest container ships
- can handle large numbers of cars
- can handle bulk cargo, such as oil, petroleum, natural gas, cement and steel products
- can supply fuel, water, food and other provisions for ships
- can repair and maintain ships.

Traders from all around the world send their goods to Singapore. From there the goods are sent on to other parts of the world. On any given day, there are about 1000 ships in the port, and a ship sails in or out of the harbour every few minutes. They connect Singapore to over 600 ports in 123 countries, spread across 6 continents. In 2012, more than 120 000 ships arrived at the port, over half of which were container ships. They carried nearly 500 million tonnes of cargo and over 1 million cruise-ship passengers.

The port of Singapore handles over 1 million cruise-ship passengers each year.

Transport links

Singapore has a modern system of roads and motorways. It also has an underground railway, with over 40 stations that link the city to its **suburbs**. The island is joined to Malaysia by a **causeway** and a bridge. There is also a ferry between Singapore and Malaysia. Singapore's airport, like its port, is one of the busiest in the world. It has flights to more than 200 other airports.

Did you know?

Singapore is the second most densely populated country in the world after Monaco in Europe. About 5.4 million people are crowded into 716 square kilometres of land. That is 7669 people for every square kilometre.

Activities

1   Write four reasons why Singapore is in a good location for a port.

2   Ferries travel between neighbouring islands and Singapore.

   a   Use travel guides, the Internet or an atlas to find as many ferry ports around the world as you can.

   b   Find out where each ferry travels to.

   c   Label these **routes** on an outline map of the world.

   d   What is the longest ferry crossing you can find?

Rotterdam, gateway to Europe

Rotterdam is the largest city in the Netherlands, and the biggest port in Europe, even though it lies 30 kilometres from the North Sea.

A map of the Netherlands.

The port area of Rotterdam.

The development of Rotterdam

Rotterdam began to develop more than 600 years ago. It started when a small group of fishermen built their huts at a marshy place where the River Maas was joined by a small stream called the Rotte. In 1340 the fishermen dug a **canal** so that they could send the fish they had caught to other villages inland. In return, the inland villages sent wheat, cheese and other farm produce to be sold in the fishing village. They sent these farm products by boat because that was the only safe way to travel in that marshy landscape. Trade grew and so did the fishing village. Eventually it became the city of Rotterdam.

The New Waterway

In 1872, a deep-water **channel**, known as the New Waterway, was dug between Rotterdam and the North Sea. The New Waterway bypassed the narrow, winding course of the River Rhine and allowed ocean-going ships to reach Rotterdam. Cargo could be unloaded at Rotterdam and put on **barges** which could take it up the River Rhine to Germany, France and Switzerland. At the same time, goods from those countries could be carried down the Rhine to Rotterdam. From there, they were exported all over the world. Trade increased greatly as a result of the New Waterway.

Rotterdam is also a focus for a network of roads and motorways, railway lines and oil pipelines that fan out all over the Netherlands and neighbouring countries. They provide more ways of distributing cargo.

The port of Rotterdam

The port of Rotterdam now stretches all the way to the sea, making it more than 40 kilometres long. Docks, warehouses, stores of coal and metal ores, flour mills, factories, shipyards and oil refineries line the banks of the waterway.

The part of the port nearest the sea is now called Europoort. The word 'poort' in Dutch means gateway. When Europoort was built in 1958, it was seen as 'the Gateway to Europe'. Europoort has expanded greatly, and can now handle the world's largest bulk carriers of oil, coal, metal ores, chemicals and grain, as well as large container ships and roll-on roll-off ships and ferries.

Loading barges from a container ship at Europoort.

Activities

1 Use an atlas and reference books or the Internet to find out:

   a the names of two countries that border the Netherlands

   b the names of three large cities in the Netherlands

   c the names of two rivers that reach the sea in the Netherlands

   d the name of the sea they flow into

   e what the flag of the Netherlands is like, then draw it

   f the language spoken in the Netherlands

   g the population of the Netherlands.

2 Collect pictures and items connected with the Netherlands. Here are some suggestions: coins, postage stamps, photographs, labels from cheese packets, garden plant packets, travel tickets.

5 Wind

Using the wind

The air around the Earth is always moving. Wind is air moving from one place to another. We use the wind's energy in several ways, but very strong winds can cause serious damage, and sometimes people and animals are killed.

How are winds formed?

Winds blow because the Sun warms up different parts of the sea and land by different amounts. The warmer parts of the Earth's surface warm the air above them. The warm air is lighter than the surrounding air, so it rises. In other places the air is cooled and becomes heavier, so it sinks. The wind blows because cold air moves to replace the warmer air that has risen.

Sea breezes

Winds often start blowing near the sea. Sometimes on a hot day you can feel a cool breeze blowing from the sea towards the beach. On a hot day, the land warms up more quickly than the sea. Air above the warm land rises, so cooler air from over the sea moves in to take its place, causing wind. At night the wind blows in the opposite direction because the land cools more quickly than the sea.

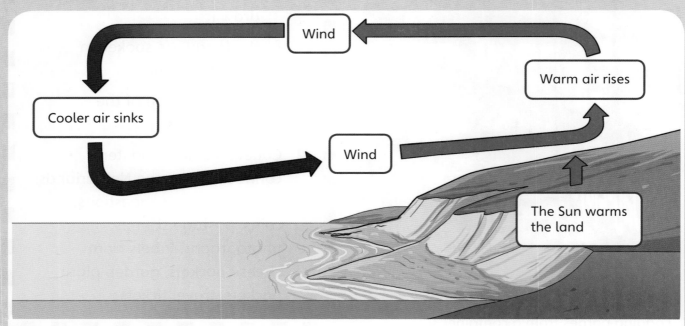

How sea breezes are formed.

The power of the wind

People have used the power of the wind for thousands of years. Long ago, people in China flew kites on a windy day to frighten their enemies. For many years, wind was used to push large sailing boats and ships around the world for exploration and trade. We still use the wind for small sailing boats today.

The earliest windmills were used to pump water or to grind corn into flour.

There are still some of these old windmills around today. Modern windmills are called wind **turbines** or wind **generators**. They have large blades, like propellers, and when these turn they drive generators that produce electricity. Unlike coal, oil and natural gas, the wind is a source of energy which does not pollute the air and which will never run out. The wind is one kind of **renewable energy**.

Sailing boats are pushed along by the wind.

Wind pumps are used to lift water from wells and drainage ditches.

Wind turbines change the energy of the wind into electricity, without polluting the air.

Activities

1 a Collect or draw pictures showing the good and bad effects of the wind.

   b Write a description of what each picture shows.

2 a Write a list of all the words you can think of that describe wind.

   b Write a list of words that describe temperature.

   c Write a list of words that describe rainfall.

   d Which is the longest list?

Wind direction and strength

We describe the wind by talking about its strength or speed, and the direction in which it is blowing.

Wind direction

We can see the direction the wind is blowing by looking at a weather vane on a tall building or a flag on a flagpole. We name a wind by the compass direction from which it blows. So a north wind is one that blows from the north towards the south. Where does a south wind blow from?

Airports, airfields and **ports** often have a windsock. This long cloth tube shows both the wind direction and the wind speed. The higher the sock lifts away from the pole, the stronger the wind.

Prevailing winds

As well as local and seasonal winds, there are certain winds that almost always blow in the same direction. These are called **prevailing winds**. At certain places on the Earth, there are belts of high air pressure and low air pressure. The air pressure is low around the Equator, for example, because the Earth's surface is very hot there, and heat makes the air expand and rise, so lowering its pressure.

There are three main belts of prevailing winds on each side of the Equator. They are called the trade winds, the Westerlies and the Polar Easterlies. They do not blow directly north–south between the pressure belts, however. This is because the spin of the Earth causes these prevailing winds to spin to the side. The diagram below shows the position of the three main belts of prevailing winds.

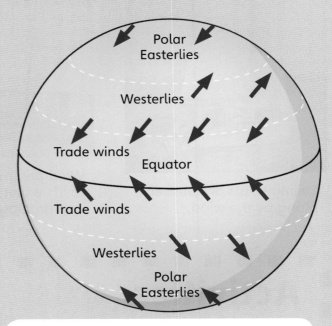

The main prevailing winds of the world.

The Beaufort scale

Wind speed is a measure of the strength of the wind. It shows how fast the air is moving.

We can estimate the speed of the wind by the effect it has on smoke, flags and tree branches, for example. Weather forecasters often use a scale worked out by Admiral Sir Francis Beaufort in 1805 for measuring wind speed at sea. His scale has been altered for use on land. The **Beaufort** **scale** goes from 0 which is dead calm with no wind at all, to 12 and over, which is a hurricane. Look at the picture below that describes the Beaufort scale.

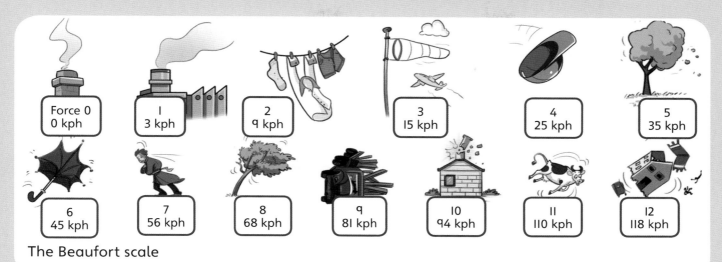

| Force 0 0 kph | 1 3 kph | 2 9 kph | 3 15 kph | 4 25 kph | 5 35 kph |
| 6 45 kph | 7 56 kph | 8 68 kph | 9 81 kph | 10 94 kph | 11 110 kph | 12 118 kph |

The Beaufort scale

How can we measure wind speed?

Sometimes it is important to describe wind speeds accurately, especially for the safety of boats and ships, aircraft and oil rigs. The scientific instrument for measuring wind speed is called an anemometer. An anemometer has four little cups that turn in the wind. The cups are attached by a shaft to a meter, rather like a car speedometer, or to a scale. The harder the wind blows, the faster the cups of the anemometer turn, and the higher the reading on the meter.

Can you see the anemometer at this weather station?

Activities

1 The prevailing winds over much of Europe come mainly from the west, especially in winter.

   a Look at a map of Europe in an atlas or on the Internet.

   b Why do you think the prevailing winds often bring wet weather?

2 Using a short stick, some tissue paper and some sticky tape, design and make a device that will show you the wind strength and direction.

Hurricanes

Hurricanes are powerful swirling storms found in the tropical parts of the Atlantic Ocean. The winds in a hurricane are very fast – over 120 kilometres an hour. Hurricanes around the Indian Ocean and Oceania are called tropical cyclones, while hurricanes in the Pacific Ocean are called typhoons.

How do hurricanes form?

Hurricanes usually begin over tropical parts of the world's oceans where the temperature is more than 27°C. They form when the air is much warmer than the ocean's surface. As this air starts swirling and moving towards the land, the storm picks up huge amounts of energy and **water vapour**. When the wind speed reaches 120 kilometres an hour a storm is called a hurricane.

Inside a hurricane

A hurricane can be as much as 900 kilometres across and 10 kilometres high. Inside it, the mass of wind and clouds spiral upwards. At the centre or 'eye' of the hurricane the skies are clear, the temperatures high and the air fairly calm.

The strongest winds, with speeds of up to 350 kilometres an hour, occur immediately around the eye of the hurricane. Hurricanes also produce very large amounts of rain – between 300 and 600 millimetres. It takes several days for a hurricane to travel from the ocean where it was formed to the land.

Hurricane names

Weather scientists give hurricanes names. Since 1978 the scientists have drawn up a list of alternate boys' and girls' names in alphabetical order. Each time a new hurricane is discovered, it is given the next name on the list. Names can be used again after six years, but the names of especially severe storms are never used again.

Hurricane damage

As it crosses the ocean, the hurricane pushes up huge waves. The water level near or under a hurricane can be over 5 metres higher than the water level in the ocean around it. When these waves reach land, they can cause serious flood damage and loss of life. Boats are lifted out of harbours, flung over sea walls and smashed to pieces far inland.

Eye of the hurricane

A satellite photograph of Hurricane Irene approaching Florida, United States in August 2011. At the 'eye' of the hurricane the skies are clear.

The hurricane winds uproot trees, destroy buildings, lift trucks off the road and break electricity and telephone poles.

Hurricane damage occurs mainly on the coast and islands. A hurricane usually weakens and dies out fairly quickly once away from the ocean because it gets its energy and moisture from the sea. On average, a hurricane lasts for a week or two.

Hurricane Sandy, in October 2012, was the largest Atlantic hurricane ever recorded. It had a diameter of 1800 kilometres. It killed at least 286 people as it moved across 7 countries. This shows some of the damage caused by Hurricane Sandy in Jamaica.

Did you know?

More than 9 million people were affected by Typhoon Haiyan, which slammed into the Philippines in November 2013, causing catastrophic damage and killing at least 5200 people.

Activities

1 a Collect newspaper cuttings or Internet news articles describing the damaging effects of the wind in various parts of the world.

   b Use the Internet to find out more details of these events.

2 When a hurricane is approaching, people are warned to take cover and to be prepared to stay where they are for up to three days.

   a Write a list of the things you would need if a hurricane was approaching your area.

   b Write them in order of importance.

3 a Imagine that you are living in an area that has been hit by a hurricane.

   b Write an account of what has happened to you, your family and your home.

   c Describe what you were thinking when the hurricane was at its worst.

   d Describe the scene that met you when you went out after the hurricane had passed by.

Tornadoes

What is a tornado? Tornadoes are small, very powerful **whirlwinds**. They form suddenly into a rapidly twisting funnel of air and cloud that stretches from the bottom of a thundercloud to the ground.

How do tornadoes form?

A tornado starts deep inside a thundercloud, when a column of rapidly rising warm air is set spinning by the strong winds blowing through the top of the cloud. Like a hurricane, a tornado is powered by rising **humid air**. However, a tornado is narrower, faster and more violent than a hurricane and it destroys everything in its path. Tornadoes often occur in small groups and, unlike hurricanes, are often found far inland. Although a tornado travels across the land at speeds of only 30 to 65 kilometres an hour, the wind speeds inside the tornado can reach 800 kilometres an hour. A tornado may measure anything from just a few metres to more than 100 metres across. It can travel for more than 200 kilometres before it uses up all its energy. Occasionally a tornado lasts for several hours, but most last for only a few minutes.

Although a tornado affects only a small area, it can cause more damage than any other kind of **weather**.

Where do tornadoes occur?

Although tornadoes can develop in most parts of the world, they are most common in the United States, where they are often called 'twisters'. On average, about 1200 tornadoes occur there every year. Tornadoes also occur regularly in parts of Canada, Argentina, China, Australia,

A large tornado destroys everything in its path. It sucks up **sand**, cars, buildings and even people and animals, like a gigantic vacuum cleaner.

South-West Asia and Europe. They can occur anywhere where there are thunderstorms.

Waterspouts

If a tornado forms over a lake or the sea it is called a waterspout. The funnel of a waterspout looks white because it contains tiny droplets of water that have cooled and **condensed** from the whirling funnel of air.

Waterspouts usually last only a few minutes. Although they are less powerful than land tornadoes, some waterspouts have wrecked boats, jetties and houses along the coast.

A waterspout over the English Channel between England and France.

Did you know?

The worst tornado ever recorded hit Missouri in the United States in March 1925. It lasted three and a half hours. It destroyed 4 towns and killed 695 people.

Activities

1 Create a spreadsheet to compare hurricanes and tornadoes. Include the following headings in your spreadsheet:

- how they form
- where they form
- their size
- their destructiveness
- any other similarities and differences you can find.

2 a Collect pictures, newspaper cuttings and Internet articles showing the path of destruction caused by a tornado.

  b If you can, plot the **route** of the tornado on a map.

3 a Look on the Internet or in reference books to find out about waterspouts.

  b Draw a diagram to show how a waterspout is formed.

Map of the World

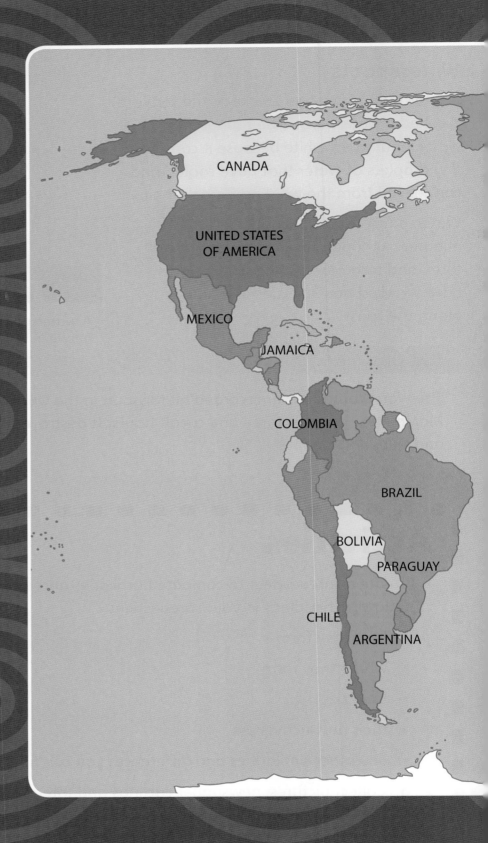

CANADA

UNITED STATES
OF AMERICA

MEXICO

JAMAICA

COLOMBIA

BRAZIL

BOLIVIA

PARAGUAY

CHILE

ARGENTINA

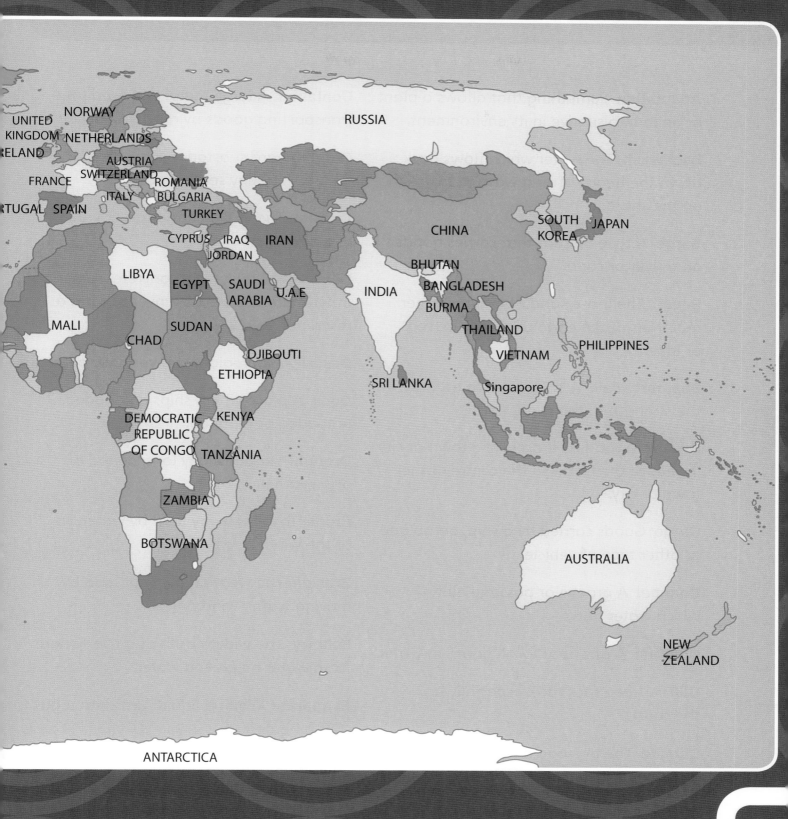

UNITED
KINGDOM
NORWAY
NETHERLANDS
RELAND
AUSTRIA
SWITZERLAND
FRANCE
ITALY
ROMANIA
BULGARIA
RTUGAL SPAIN
TURKEY
CYPRUS IRAQ IRAN
JORDAN
LIBYA
EGYPT
SAUDI
ARABIA
U.A.E
MALI
CHAD
SUDAN
DJIBOUTI
ETHIOPIA
DEMOCRATIC
REPUBLIC
OF CONGO
KENYA
TANZANIA
ZAMBIA
BOTSWANA
RUSSIA
CHINA
SOUTH
KOREA
JAPAN
BHUTAN
INDIA
BANGLADESH
BURMA
THAILAND
VIETNAM
PHILIPPINES
SRI LANKA
Singapore
AUSTRALIA
NEW
ZEALAND
ANTARCTICA

Glossary

Adaptation Something that allows a plant or animal to survive in its environment.

Backwash The water which flows back down the beach after a wave has broken on the shore.

Barge A type of boat that carries goods on a river or canal.

Beach The loose sand, shingle or other materials at the edge of an ocean, sea or lake.

Beaufort scale A scale used to estimate the strength of the wind.

Bypass A road built around a town or city to reduce the traffic in the centre of the town or city.

Cargo Goods carried by a ship, aeroplane or other type of vehicle.

Channel A groove or passage along which water flows

Chlorine A gas used to kill germs in water.

Cliff A steep rock face, especially on the coast.

Climate The average or typical weather of a region of the Earth throughout the year.

Coast Where the land meets the sea.

Condense When a gas turns into a liquid as it is cooled.

Container A large metal box used for transporting goods by road, rail or sea.

Coral A hard substance made from the skeletons of tiny sea animals.

Current The movement of air or water in a particular direction.

Dam A wall built from concrete, soil or other materials across a river to form a reservoir.

Desert A dry region with very few plants.

Dock A place where ships are loaded, unloaded or repaired.

Drought An unusually long period of dry weather.

Erode To wear away land by water, ice or the wind.

Erosion The wearing away of land by moving water, wind or ice.

Estuary The wide mouth of a river where fresh water meets sea water.

Evaporate When a liquid turns into a gas as it is heated.

Famine A time when there is not enough food for the people of an area.

Filter beds Layers of sand and gravel that are used to separate insoluble particles from water in a water treatment works.

Glacier A slow-moving river of ice.

Global warming A gradual increase in the average temperature of the Earth's climate caused by certain kinds of air pollution.

Groundwater Water in the rocks below the Earth's surface.

Harbour A place where ships and boats can shelter or unload.

Hardwood A type of wood that is strong. Oak, mahogany, teak and rosewood are hardwoods.

Humid air Air that contains a large amount of water vapour.

Hurricane A powerful swirling storm found in tropical parts of the Atlantic Ocean.

Industry The making of things in factories or workshops.

Irrigation The taking of water to the land so that crops grow well.

Lake A large hollow in the land which is filled with water.

Latrine A very basic type of toilet.

Monsoon A strong wind in and around the Indian Ocean which brings heavy rain in summer.

Motorway A wide road built for fast-moving cars, trucks and coaches.

Nomad A person who lives and works on the move from one place to another.

Non-renewable fuels Fuels that cannot be replaced once used.

Oasis A place where water can be found in a desert.

Pollute To make dirty.

Pollution When substances such as air, water or the soil are spoiled or made dirty.

Port A place where ships stop to load and unload their cargo.

Power station A large building where electricity is produced.

Prevailing wind A wind that almost always blows from the same direction.

Rainforest A forest that grows in a place that is hot and rainy all the year round.

Reservoir A lake built to store water.

Resort A place where people go for a holiday.

Routeway A way of getting from one place to another.

Sand dune A hill formed by wind-blown sand.

Settlement A place, such as a village, town or city, where people live.

Sewage The waste material and liquid from houses and factories, carried away by drains and sewers.

Spring A place where water flows out of the ground.

Stack A pillar of rock in the sea near cliffs.

Tides The rise and fall of the level of the oceans and seas twice a day.

Tornado (or twister) A very violent whirlwind.

Transport Ways of taking people or goods from one place to another.

Tropics The regions near the Equator that have a hot climate all the year round.

Turbine A type of fan that is turned by steam, water pressure or the wind.

Typhoon A weather system that brings very powerful winds and torrential rain, and causes widespread destruction.

Waterspout A tornado over the sea.

Water treatment plant The place where water is cleaned ready to be supplied.

Water vapour The gas that forms when water is heated.

Wave A regular movement of the surface of water caused by the wind.

Weather How hot or cold, wet or dry, moving or still the air is at a particular time.

Well A hole that has been dug or drilled into the ground to reach underground water, oil or gas.

Whirlwind A strong wind that blows round and round.